I0796312

WHAT IS METEOROLOGY?

Meteorology is the study of the weather, including the different types, the causes, and the ways to predict it.

The scientists who study the weather are called **METEOROLOGISTS.**

Words that are tricky to understand are in **bold**. Find out what they mean in the glossary.

Words that are difficult to say are in *italics.* Find out how to say them at the back of the book.

CAN PREDICTING THE WEATHER SAVE LIVES?

DISCOVER THE SCIENCE BEHIND **METEOROLOGY**
(mee-tee-uh-ROL-uh-jee)

Written by Eliza Jeffery
Illustrated by Daniel Limon

It is very important to know what the weather will be like today and in the future. A fishing boat crew spends many days at sea, catching fish. The crew relies on weather warnings. As big storms develop, they need to return home to safety.

Changes in our weather caused by **climate change** can be devastating. Arctic sea ice that polar bears depend on as a platform for hunting is melting and they are going hungry.

We need the right amount of rain and sunshine to grow food. Farmers rely on **forecasts** to find out when to plant and harvest their crops.

The people who forecast the weather are *meteorologists*. These scientists need to gather information to do their jobs. They use **satellites, radar,** and **weather stations** to do this. But they have smaller tools to help them too!

Anemometer
Wind speed

Snow gauge
Snow

Barometer
Air pressure

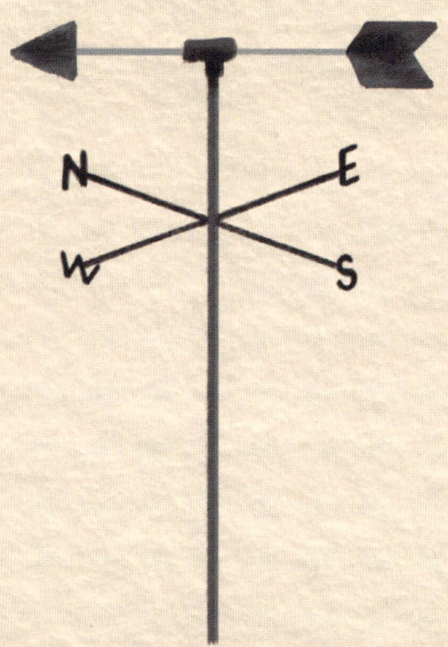

Wind vane
Wind direction

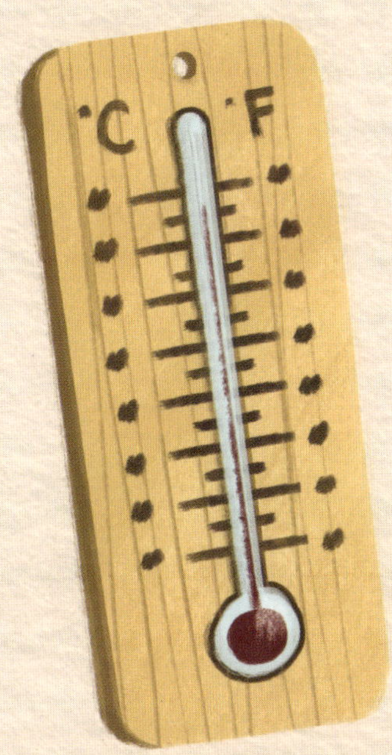

Thermometer
Temperature

Wind sock
Wind direction

Rain gauge
Rain

Weather balloon
Wind, temperature, and humidity

Some places around the world have extreme weather. In the Pacific Ocean region, **typhoon** season brings storms like **cyclones** and **monsoons.** Warnings and forecasts are lifesavers because they help people prepare for strong winds, heavy rain, and flooding.

Hurricanes are another very dangerous type of storm. They grow from a mix of warm ocean waters and thunderstorms. Extreme winds can make hurricanes stronger and sometimes push them from the sea onto land...

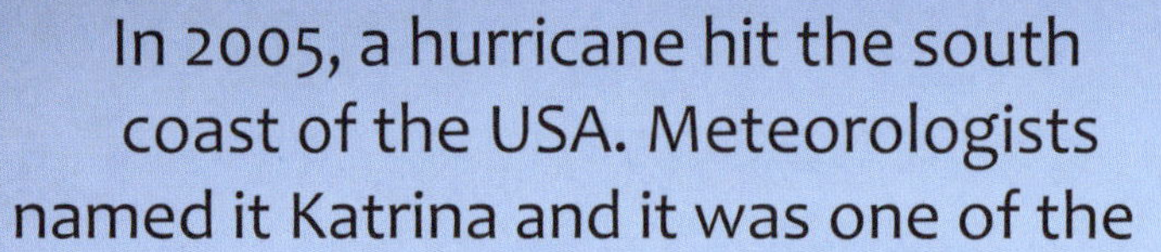

In 2005, a hurricane hit the south coast of the USA. Meteorologists named it Katrina and it was one of the

MOST DESTRUCTIVE HURRICANES EVER RECORDED!

This deadly storm caused $160 billion (£126 billion) in damage to buildings and homes, and sadly many people lost their lives.

Emergency teams, military, and local volunteers were all called in. Meanwhile, meteorologists were busy working behind the scenes, tracking Hurricane Katrina's progress.

Paramedics and doctors

Fire services

Volunteers

Coast guards

Police

Meteorologists

Meteorologists watched the hurricane closely. They were able to get a perfect view of the storm from space using satellites. They could see the size of the hurricane and where it was headed. And they learned that it was getting bigger as it reached land...

Forecasts and warnings were sent to people in the **impact zones.** Some were told to stock up on food, and some had to **evacuate.** Hurricane Katrina just kept getting

STRONGER AND STRONGER!

Day and night, meteorologists kept tracking the path of the hurricane. This way, they could predict floods and the places that would be damaged. Then emergency response teams could move people to safety and help where it was needed.

Meteorologists always watch the weather closely. Sometimes, big events like volcanoes erupting fill the air with ash.

This can make it hard to breathe and cause big changes to the weather

Volcanic ash can be carried long distances by the wind, covering surfaces for miles around. This creates unhealthy **toxins** and a rapid change in temperature.

Meteorologists are always on the lookout for changes in weather, even the kinds caused by natural disasters.

CLOSED DUE TO
BAD WE THER

Rescues can’t happen without the all-clear from meteorologists. Mountain rescuers and firefighters rely on detailed forecasts for their own safety.

Meteorologists can use this sort of information to help in other situations. Some sports depend on weather forecasts to know the best times to play or to plan events.

After Hurricane Katrina, meteorologists continued to study the storm to help them understand weather patterns better. They want to be able to predict severe weather like cyclones, typhoons, and hurricanes to keep everyone safe and help save lives across the world. They are the

UNSUNG HEROES OF SCIENCE!

Real-life

EXTREME WEATHER

Now we know all about Hurricane Katrina, but what other extreme weathe events have happened that have changed how we predict the weather?

Indian RED RAIN!

In 2001, the rainfall in Kerala, India, was blood red! This happened when high leve of red dust got mixed with the rain.

Californian WILDFIRES!

A wildfire is a type of fire that is uncontrollable and unplanned in a natural **habitat**. The 2018 wildfire in California is one of the largest and most destructive wildfires in history!

Iranian BLIZZARD!

In 1972, a deadly blizzard swept across Iran, covering over 200 villages in thick snow. This was the worst blizzard ever recorded there!

Pakistani FLOODS!

In 2022, Pakistan recorded one of the heaviest floods in its history. This heavy rainfall caused flooding across one third of the country!

Mount Tambora ERUPTION!

In 1815, a large volcano called Mount Tambora erupted in Indonesia. It was the most powerful volcanic eruption in recorded human history!

Wacky WEATHER FACTS

So, predicting the weather really can save lives, but what else is there to know? What else have meteorologists discovered about the weather?

CRICKETS CAN TELL THE TEMPERATURE!

Meteorologists aren't the only ones who gather information about the weather. You can tell the temperature by how often a cricket chirps!

THE LARGEST HAILSTONE RECORDED WAS THE SIZE OF A FOOTBALL!

Hailstone showers are when small lumps of ice fall from the sky like rain. They are usually very small… but some can be bigger!

WATER ON EARTH IS ALWAYS ON THE MOVE.

Rain falling today may have been water in a distant ocean days before! Water is constantly moving because of the force of **gravity** and energy from sunlight.

WILDFIRES CAN CREATE TORNADOES MADE OF FIRE!

When a wildfire picks up high-speed winds, it often creates a spinning sensation with the flames. These are called fire whirls and can be deadly!

RAINBOWS ARE ACTUALLY FULL CIRCLES.

We may only see rainbows as a half-circle from land, but from the sky, people have often spotted rainbows in full circular form!

GLOSSARY

Climate change – a change in the weather conditions over a long time.

Cyclones – powerful storms that spin at high speeds. *Need help saying this? Look below!*

Evacuate – to move people from an area where they are in danger to a safer place.

Forecasts – guesses or estimates about something that will happen in the future.

Gravity – an invisible force that pulls things together. Gravity keeps your feet on the ground and keeps Earth spinning around the Sun.

Habitat – the place where animals or plants live.

Impact zones – areas where a storm, a wave, or another event will strike.

Monsoons – seasonal winds which last for several months across a whole region. It often comes with heavy rain.

Radar – a machine that uses radio waves to sense objects and natural structures like clouds. *Need help saying this? Look below!*

Satellites – things in space that go around a planet like Earth. Human-made satellites can take pictures of the Earth's surface. *Need help saying this? Look below!*

Toxins – the substances created by plants and animals that are dangerous to humans.

Typhoon – a circular storm that forms over warm oceans.

Weather stations – places where wind speed, wind direction, temperature, and air pressure are measured.

HOW DO I SAY?

Cyclones
SY-klohns

Hurricanes
HUHR-uh-kanes

Meteorologists
mee-tee-uh-ROL-uh-jists

Meteorology
mee-tee-uh-ROL-uh-jee

Radar
RAY-dahr

Satellites
SAT-uh-lyts

THE BIG QUESTIONS ANSWERED

This is more than just a series of books; it is a complete resource. Accompanying each book is a variety of FREE material to engage curious kids with science.

www.thebigquestionsanswered.com

Use the QR code to visit the website, download free resources, and discover other books in the series.

On the website, find out incredible things about meteorologists, including what they do, some of their greatest discoveries, and the people who have made a difference in this field of science.

The material is also available for home or classroom use, supporting all the information in this book.

Teachers' & Parents' Resources
With discussion prompts and questions, extra information, and facts around key topics.

Young Meteorologists' Activity Pack
Fun activities for wannabe weather experts, including creative writing, drawing, word searches, and much, much more.

BEETLE BOOKS

The Big Questions Answered is published by Beetle Books.
Beetle Books is an imprint of Hungry Tomato Ltd.

First published in 2024 by Hungry Tomato Ltd
F15, Old Bakery Studios, Blewetts Wharf, Malpas Road,
Truro, Cornwall, TR1 1QH, UK.

ISBN 9781835691328

A CIP catalog record for this book is available from the British Library.

With thanks to:
Editor: Millie Burdett
Editor: Holly Thornton
Senior Designer: Amy Harvey
Tim Cook for his valued contribution
The team at Beehive Illustration

Printed and bound in China.

Picture Credits:
(t = top, b = bottom, m = middle, l = left, r = right)
Shutterstock: Alaskagirl8821 32br; Andrei Armiagov 35tl; encikAn 34mr; Justkgoomn 33tr; ohenze 35mr; Philip B. Espinasse 35bl; Saigh Anees 33ml; Suzanne Tucker 34bl.